THOMAS CRANE PUBLIC LIBRARY
QUINCY MASS
CITY APPROPRIATION

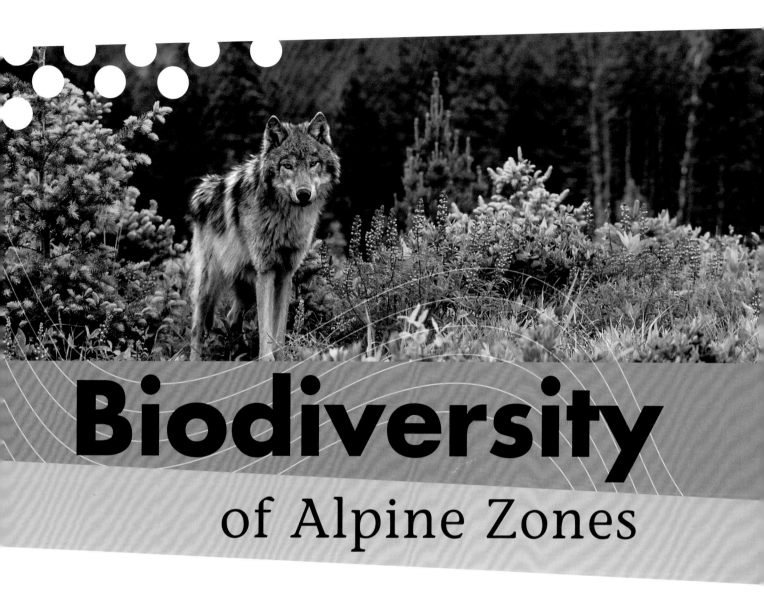

Biodiversity
of Alpine Zones

GREG PYERS

Marshall Cavendish
Benchmark
New York

This edition first published in 2012 in the United States of America by
MARSHALL CAVENDISH BENCHMARK
An imprint of Marshall Cavendish Corporation

All rights reserved.

No part of this publication may be reproduced, stored in a retrieval system or transmitted, in any form or by any means, electronic, mechanical, photocopying, recording, or otherwise, without the prior permission of the copyright owner. Request for permission should be addressed to the Publisher, Marshall Cavendish Corporation, 99 White Plains Road, Tarrytown, NY 10591. Tel: (914) 332-8888, fax: (914) 332-1888.

Website: www.marshallcavendish.us

This publication represents the opinions and views of the author based on Greg Pyer's personal experience, knowledge, and research. The information in this book serves as a general guide only. The author and publisher have used their best efforts in preparing this book and disclaim liability rising directly and indirectly from the use and application of this book.

Other Marshall Cavendish Offices:
Marshall Cavendish International (Asia) Private Limited, 1 New Industrial Road, Singapore 536196 • Marshall Cavendish International (Thailand) Co Ltd. 253 Asoke, 12th Flr, Sukhumvit 21 Road, Klongtoey Nua, Wattana, Bangkok 10110, Thailand • Marshall Cavendish (Malaysia) Sdn Bhd, Times Subang, Lot 46, Subang Hi-Tech Industrial Park, Batu Tiga, 40000 Shah Alam, Selangor Darul Ehsan, Malaysia

Marshall Cavendish is a trademark of Times Publishing Limited

All websites were available and accurate when this book was sent to press.

Library of Congress Cataloging-in-Publication Data

Pyers, Greg.
 Biodiversity of alpine zones / Greg Pyers.
   p. cm. — (Biodiversity)
 Includes index.
 Summary: "Discusses the variety of living things in the ecosystem of alpine regions"
 —Provided by publisher.
 ISBN 978-1-60870-528-3
 1. Mountain biodiversity—Juvenile literature. 2. Mountain ecology—Juvenile literature.
 3. Endangered ecosystems—Juvenile literature. I. Title.
 QH87.P94 2012
 577.5/3—dc22
                    2010037460

First published in 2011 by
MACMILLAN EDUCATION AUSTRALIA PTY LTD
15–19 Claremont Street, South Yarra 3141

Visit our website at www.macmillan.com.au or go directly to www.macmillanlibrary.com.au

Associated companies and representatives throughout the world.

Copyright © Macmillan Publishers Australia 2011

Publisher: Carmel Heron
Commissioning Editor: Niki Horin
Managing Editor: Vanessa Lanaway
Editor: Georgina Garner
Proofreader: Tim Clarke
Designer: Kerri Wilson
Page layout: Raul Diche
Photo researcher: Wendy Duncan (management: Debbie Gallagher)
Illustrator: Richard Morden
Production Controller: Vanessa Johnson

Printed in China

Acknowledgments
The author and publisher are grateful to the following for permission to reproduce copyright material:

Front cover photograph: Gray wolf in Rocky Mountains, USA courtesy of photolibrary/J-L. Klein & M-L. Hubert.
Back cover photographs courtesy of Shutterstock/Zoran Karapancev (mountain goat), /photostockar (edelweiss).

Photographs courtesy of:
ANTphoto.com.au/Mrs JM Soper, 23, /Stock Connection/Mountain Light/Galen Rowell, 19; Auscape/BIOS/Alain Dragesco-Joffe, 9; Corbis/Anup Shah, 18; Fairfaxphotos/Andrew Taylor, 20; FLPA/Bob Gibbons, 4; Getty Images/The Image Bank/picturegarden, 22, /National Geographic/Nicholas Devore III, 24; istockphoto/Dan Eckert, 11; National Geographic Stock/Alaska Stock Images, 7; Nature Picture Library/Gavin Maxwell, 15, /T J Rich, 25; photolibrary/Alamy/Arco Images GmbH, 13, /Alamy/blickwinkel, 17, /OSF/Emanuele Biggi, 27; Shutterstock/Chelsea, 16, /douwevrs, 21, /Morag Fleming, 10, /A Ryser, 28, /David Thyberg, 14. Background and design images used throughout courtesy of Shutterstock/photostockar (edelweiss), /fotoret (snowdrops).

While every care has been taken to trace and acknowledge copyright, the publisher tenders their apologies for any accidental infringement where copyright has proved untraceable. They would be pleased to come to a suitable arrangement with the rightful owner in each case.

Please note
At the time of printing, the Internet addresses appearing in this book were correct. Owing to the dynamic nature of the Internet, however, we cannot guarantee that all these addresses will remain correct.

1 3 5 6 4 2

Contents

What Is Biodiversity?	4
Why Is Biodiversity Important?	6
Alpine Zones of the World	8
Alpine Biodiversity	10
Alpine Ecosystems	12
Threats to Alpine Zones	14
Biodiversity Threat: Tourism	16
Biodiversity Threat: Grazing and Poaching	18
Biodiversity Threat: Invasive Species	20
Biodiversity Threat: Climate Change	22
Alpine Conservation	24
Case Study: The Alps	26
What Is the Future of Alpine Zones?	30
Glossary	31
Index	32

Glossary Words

When a word is printed in **bold**, you can look up its meaning in the Glossary on page 31.

What Is Biodiversity?

Biodiversity, or biological diversity, describes the variety of living things in a particular place, in a particular **ecosystem**, or across the entire Earth.

Measuring Biodiversity

The biodiversity of a particular area is measured on three levels:
- **species** diversity, which is the number and variety of species in the area.
- genetic diversity, which is the variety of **genes** each species has. Genes determine the characteristics of different living things. A variety of genes within a species enables it to **adapt** to changes in its environment.
- ecosystem diversity, which is the variety of **habitats** in the area. A diverse ecosystem has many habitats within it.

A ground squirrel feeds on a diverse range of wild plants in an alpine meadow habitat. The squirrel and plants are all part of alpine biodiversity.

Species Diversity

Some habitats, such as rain forests, have very high biodiversity. More than fifty species of ants might be found in just 11 square feet (1 square meter) of the Amazon Rain Forest floor, in South America. The same area of ground in an English woodland, however, might have just one species of ant. Alpine habitats have a large variety of small flowering plant species, but in forest habitats, there are very few flowering plants.

Habitats and Ecosystems

Alpine zones are habitats, which are places where animals and plants live. Within an alpine zone, there are many smaller habitats, sometimes called microhabitats. Some alpine microhabitats are rock crevices, alpine bogs, and cliffs. Different kinds of **organisms** live in these places. The animals, plants, other living things, nonliving things, and all the ways they affect each other make up an alpine ecosystem.

Biodiversity Under Threat

The variety of species on Earth is under threat. There are somewhere between 5 million and 30 million species on Earth. Most of these species are very small and hard to find, so only about 1.75 million of these species have been described and named. These are called known species.

Scientists estimate that as many as fifty species become **extinct** every day. Extinction is a natural process, but human activities have sped up the rate of extinction by up to one thousand times.

Did You Know?

About 95 percent of all known animal species are invertebrates, which are animals without backbones, such as insect, worm, spider, and mollusk species. Vertebrates, which are animals with backbones, make up the remaining 5 percent.

Known Species of Organisms on Earth

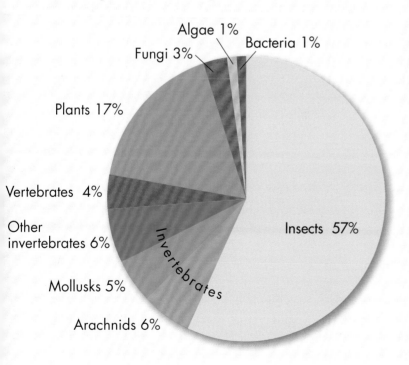

- Algae 1%
- Bacteria 1%
- Fungi 3%
- Plants 17%
- Vertebrates 4%
- Other invertebrates 6%
- Mollusks 5%
- Arachnids 6%
- Insects 57%
- Invertebrates

Approximate Numbers of Known Vertebrate Species

ANIMAL GROUP	KNOWN SPECIES
Fish	31,000
Birds	10,000
Reptiles	8,800
Amphibians	6,500
Mammals	5,500

The known species of organisms on Earth can be divided into bacteria, algae, fungi, plant, and animal species. Animal species are further divided into vertebrates and invertebrates.

Why Is Biodiversity Important?

Biodiversity is important for many reasons. The diverse organisms in an ecosystem take part in natural processes essential to the survival of all living things. Biodiversity produces food and medicine. It is also important to people's quality of life.

Natural Processes

Humans are part of many ecosystems. Our survival depends on the natural processes that go on in these ecosystems. Through natural processes, air and water are cleaned, waste is decomposed, **nutrients** are recycled, and disease is kept under control. Natural processes depend on the organisms that live in the soil, on the plants that produce oxygen and absorb **carbon dioxide**, and on the organisms that break down dead plants and animals. When species of organisms become extinct, natural processes may stop working.

Food

We depend on biodiversity for our food. The world's major food plants have all been bred from plants in the wild. Wild plants are the source of genes for breeding new varieties of plants. New varieties may be bred to be resistant to disease or to be grown in a warmer **climate** or in poor soil. When wild plants become extinct, their useful genes are lost.

Medicine

About 40 percent of all prescription drugs come from chemicals that have been extracted from plants. Scientists discover new, useful plant chemicals every year. One scientific study found that 62 percent of Himalayan alpine plants in Tibet are used by local people for traditional medicines or for other uses. Many of these plants could be important in the making of new medicines for people all around the world.

Did You Know?

Across the world, barley is an important grain crop. It was first bred 3,500 years ago from wild barley grass, which grows in the alpine zones of Tibet. Wild barley still grows in the mountains.

The sight of a mountain goat on an alpine slope can inspire wonder and imagination. This improves our quality of life.

Quality of Life

Biodiversity is important to people's quality of life. Animals and plants inspire wonder. They are part of our **heritage**. People enjoy walking in the mountains because of the biodiversity they see, such as an eagle soaring or alpine flowers growing in springtime. Without alpine biodiversity, high mountains would be empty, lifeless places of rock and ice.

Endangered Species

The kea of New Zealand is the world's only alpine parrot species. Before 1970, humans hunted and killed more than 150,000 keas because they were attacking sheep and feeding on the fat beneath the skin. The kea is now in danger of extinction. Fewer than 5,000 of these birds remain in the wild.

Alpine Zones of the World

Alpine zones are above the tree line, which is the **altitude** where it becomes too cold for trees to grow. These zones are found at the tops of mountains or in high valleys or plains. Alpine zones make up only about 3 percent of Earth's surface.

Areas Above the Tree Line

The tree line is at a different height above sea level in different parts of the world. In the mountains of northern Scandinavia, it may be about 1,000 feet (300 meters) above sea level. The tree line of Mount Kenya, near the Equator in Africa, is at about 9,850 feet (3,000 m).

If an alpine zone is at the top of a mountain, it is often like an island of high ground, surrounded by lower-lying habitats. Larger, flat areas in high valleys or plains, called plateaus, may be in the alpine zone, too.

Alpine **vegetation** is found between the tree line and the permanent snow line. Very few, if any, plants can grow above the permanent snow line, because there is no soil and it is very cold.

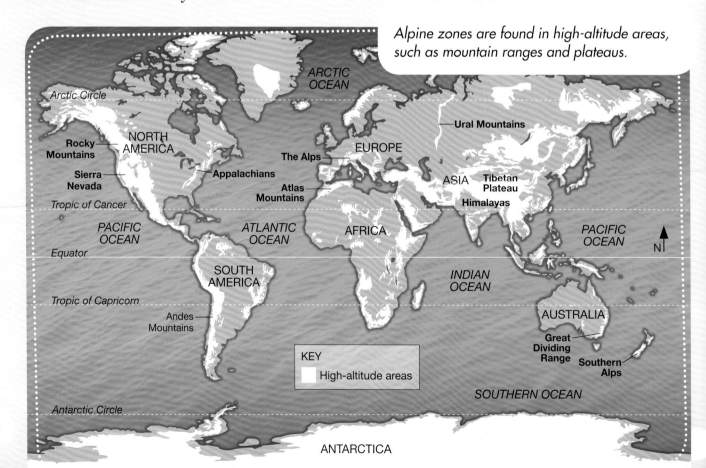

Alpine zones are found in high-altitude areas, such as mountain ranges and plateaus.

The Tibetan Plateau

The largest area of alpine zone is on the Tibetan Plateau in Asia. The plateau covers 965 square miles (2,500 sq km) and has an average elevation of 14,760 feet (4,500 m). Strong winds, cold temperatures, intense solar radiation, and low-pressure and low-oxygen air create harsh living conditions, where few animal and plant species can survive.

There are many habitat types on the Tibetan Plateau, from forests to alpine meadows. Alpine meadows are natural pastures that grow above the tree line. They are habitat for white-lipped deer, Tibetan wild ass, Tibetan gazelle, and wild sheep such as argali and bharal, or blue sheep. These animals are preyed on by carnivores such as snow leopards, wolves, and brown bears.

Tundra

Alpine habitats belong to a group of habitats called tundra. Tundra is a treeless habitat with low-growing plants such as dwarf shrubs, mosses, lichens, and grasses. Tundra is found in cold climates, such as in polar regions and on high mountains. Arctic tundra consists of vast, treeless plains that have permanently frozen soil, called permafrost, about 20 inches (50 centimeters) below the ground. Alpine habitats do not usually have permafrost.

Small antelopes called Tibetan gazelles graze on the alpine plains of the Tibetan Plateau in Asia.

Alpine Biodiversity

Alpine zones have lower biodiversity than some other types of habitats, such as rain forests, but they may have many **endemic species** and many species that are rarely seen.

Endemic and Rare Species

An alpine zone often has many endemic species because its alpine habitats are separated by areas of lower altitude, so plants and animals in one alpine area cannot breed with plants and animals in another alpine area. Over time, different species **evolve** in these different alpine areas.

An alpine species may be rarely seen for a number of reasons, including:

- the species is endemic to just a small area of alpine zone
- the species is very small
- the species lives in a hard-to-reach, isolated area
- the species does not exist in large numbers.

Mountain Pygmy Possum

The mountain pygmy possum, or burramys, is a rare species that is endemic to Australia's alpine zone. This tiny marsupial is found in three small, isolated areas in Victoria and New South Wales, where it lives among boulders. Its total population is about 2,250 and its total **range** is only 2.7 square miles (7 sq km). In the summer, it feasts on bogong moths that fly to the mountains to shelter. In the winter, it **hibernates** among the boulders.

The Himalayan blue poppy is endemic to alpine zones in Tibet, but it is now grown in many gardens around the world, too.

The Sierra Nevada mountain range in California has high species diversity in its different alpine habitats. Many of its plant species are endemic to one location only.

Alpine Plants of Iran

There are 682 known species of plants in the alpine zone of Iran. About 396 of these are endemic species, some of which are very rare. For about 40 percent of the endemic species, only one specimen has ever been recorded. For another 13 percent, only two or three specimens have ever been recorded.

As in other alpine zones, the number of plant species in Iran's alpine zone declines as the altitude increases. This is because temperatures are lower at higher altitudes, and it becomes too cold for all but the hardiest plants to grow.

Alpine Plants of California

A study of alpine plants in California showed how one alpine area can be very different from another. When scientists studied the biodiversity of fourteen mountain summits, they found that about half of all species of plants were found on one summit only. Only one plant species was found across all fourteen summits.

Alpine Ecosystems

Living and nonliving things, and the **interactions** between them, make up alpine ecosystems. Living things are plants and animals. Nonliving things are the rocks, soil, and water, as well as the climate.

Food Chains and Food Webs

A very important way that different species interact is by eating or consuming other species. This transfers energy and nutrients from one organism to another. A food chain illustrates this flow of energy, by showing what eats what. Food chains are best seen in a diagram. A food web shows how many different food chains fit together.

This European alpine food web is made up of several food chains. In one food chain, grasses are eaten by Alpine marmots, which in turn are eaten by golden eagles.

Flowers and butterflies interact in alpine ecosystems. The flowers produce nectar that the butterflies feed on, and the butterflies spread pollen from flower to flower.

Other Interactions

Living things in an alpine habitat interact in other ways, too. In the brief springtime of an alpine zone, flowering plants have just a short time to produce seeds before they are covered in the winter snow. To produce seeds, pollen must first be transferred from flower to flower. It is usually transferred by insects such as butterflies. Alpine plants produce sweet nectar to attract the insects, and their bright, showy flowers advertise that they have nectar to drink. The nectar gives insects the energy to fly, mate, and lay eggs. As the insects fly from flower to flower to feed, they transfer pollen.

Seasons and Biodiversity

Many alpine zones have very distinct seasons, and these seasons affect the biodiversity of the zone. In the summer, many animals arrive to feed on the plentiful food. By the winter, most of these species have moved below the alpine zone to escape the harsh weather.

Threats to Alpine Zones

Alpine zones are less threatened by human activities than other areas, mainly because so few people live in alpine zones. However, alpine biodiversity is threatened by tourism, grazing, **poaching**, and **invasive species**. Climate change is also a major threat.

Biodiversity Hotspots

There are about thirty-four regions in the world that have been identified as biodiversity hotspots. These hotspots are areas that have a high number of endemic species and biodiversity that is still mainly intact, but this biodiversity is under severe threat from humans. Some of the world's biodiversity hotspots include alpine zones with a high number of endemic species.

The guanaco lives in the Tropical Andes, one of four biodiversity hotspots with alpine zones.

Biodiversity Hotspots That Include Alpine Zones

HOTSPOT	ALPINE BIODIVERSITY	THREATS TO ALPINE BIODIVERSITY
Himalayas	This hotspot has the highest recorded vascular plant (a plant that has veins that carry water and nutrients through its stems and leaves), called *Ermania himalayensis*, which is found on **scree** at an altitude of 20,670 feet (6,300 m). It is also home to the black-necked crane, which is the world's only alpine crane.	Overgrazing in the summer by **domestic** animals Poaching (such as poaching snow leopards for their **pelts** and wild sheep for their horns) Hunting of hoofed mammals for food Competition for grazing with domestic animals
Mountains of central Asia	Endemic mammals include Menzbier's marmot and ili pika. Grazing mammals include argali, bharal, and Siberian ibex.	Snow leopards are killed to protect domestic livestock Taking of plants for traditional medicines Increasing average temperatures due to climate change are causing the tree line to rise higher up mountains, shrinking alpine zones
Mountains of southwestern China	The black snub-nosed monkey is seen at higher altitudes than any other non-human primate—up to 14,760 feet (4,500 m) above sea level.	
Tropical Andes, South America	There are more than 800 plant species in the alpine grasslands. *Parajubaea torallyi* grows at higher altitudes than any other palm species—11,150 feet (3,400 m). Grazing animals include vicuna and guanaco.	Overgrazing by domestic animals Mining causes habitat loss and **erosion** Climate change

The black-necked crane is an alpine species that lives in the Himalayan biodiversity hotspot, where it breeds on the Tibetan Plateau in spring. The population of this threatened species is about 6,000.

BIODIVERSITY THREAT:
Tourism

Not many people live in alpine zones, but tourists visit these areas all year round. They come to see spectacular scenery and wildlife, but also to hike, ride mountain bikes, ski, and rock climb. These activities can threaten alpine biodiversity.

Negative Effects of Tourism

The negative effects of tourism on alpine biodiversity are often difficult to measure. However, the more people that enter an alpine zone, the greater the risk that vegetation will be trampled, nonnative species will be introduced, and alpine wildlife will be disturbed or killed.

Loss of Habitat and Introduced Species

Ski resorts **urbanize** alpine zones, replacing habitat with buildings and introducing noise, litter, and traffic. Roads are built so that tourists can reach ski resorts, and the cars and trucks that use the roads kill wildlife and introduce nonnative species.

Many alpine species are unable to survive in these changed environments, but introduced species may thrive. Black rats are carried to ski resorts in vehicles. They survive the cold winter living in and around ski resort buildings, and in the summer, they spread out to prey on alpine birds and mammals.

Ski resorts cause major disturbances for alpine species. When ski runs are built, vegetation is cleared and species lose their natural habitats.

Fragmented Habitats

Roads can break up an alpine habitat into separate fragments, and they can be uncrossable barriers for many animal species. Animals might be killed crossing a road, they might be scared away due to traffic noise, or a road might be fenced off and unpassable.

Species that live in a fragmented habitat might not be able to **migrate** or move to new areas to feed. When a large group of a species is broken into small, isolated groups, animals cannot look for a mate outside their small group, so inbreeding occurs. Inbreeding causes each smaller group to lose genetic diversity and its members become closely related to one another. Inbred populations may have genetic defects that reduce a species' ability to reproduce or make it vulnerable to disease.

Benefits of Tourism

Nature-based tourism, or ecotourism, takes people into alpine areas to view scenery and wildlife. Tourists who experience and appreciate the beauty of an alpine zone often become supporters of conservation and help protect alpine zones. However, this kind of tourism can still threaten alpine biodiversity if it is not well managed.

In the European Alps, deer need to migrate between their summer and winter habitats. Human developments such as farms and roads interfere with their migration.

BIODIVERSITY THREAT:
Grazing and Poaching

In some alpine areas, domestic animals such as cows and sheep have taken over the habitats of native grazing species. Other native animals, such as snow leopards, are hunted because they kill livestock or they are poached for their pelts or meat.

Grazing Domestic Animals

When graziers bring sheep, cattle and goats into the mountains, these animals compete with native grazing animals for food. Due to this competition, native grazing animals have fallen in number in many parts of their range. This means that **predators** that usually prey on these native animals are forced to hunt domestic animals instead. Many graziers try to protect their herds by shooting, trapping, or poisoning these predators.

Snow Leopards

In the alpine and subalpine areas of Asia, snow leopards hunt a range of native grazing animals, such as bharal, Siberian ibex, markhor, and Himalayan tahr. Today, there may be as few as 2,500 snow leopards across their entire range, due to a loss of their natural prey and poaching.

Ethiopian Wolves

Ethiopian wolves live at altitudes of 11,480 feet (3,500 m) or higher in the mountains of Ethiopia. Many have been hunted or driven from their habitats by ranchers or herders and their dogs. It is now one of the world's rarest mammals.

Did You Know?

Farm dogs carry the diseases rabies and distemper, which are deadly to Ethiopian wolves. Programs such as the Ethiopian Wolf Conservation Programme vaccinate dogs in certain regions so that they cannot transmit diseases to wolves.

There may be as few as 500 Ethiopian wolves in the wild. The wolves are endangered due to hunting by graziers and introduced diseases.

Poaching Native Species

Snow leopards are poached for their pelts, teeth, and claws. A snow leopard pelt may be sold for thousands of dollars. This is a great temptation for people whose incomes may be as low as a few dollars a day.

Alpine grazing animals are also poached, mainly for their meat. In Pakistan, the endangered markhor is hunted for its large, spiral horns.

A family in Pakistan, in Asia, displays a poached snow leopard pelt for sale. The pelt will earn them a large amount of money.

Grazing Animals and Clean Water

Up to 40,000 cattle are permitted to graze every summer in the alpine zone of the Sierra Nevada mountains in California. Since the 1880s, it has been known that grazing leads to reduced water quality, because animal dung pollutes the water. However, it is now known that grazing also changes alpine vegetation, because some plant species are eaten more than others. Cattle grazing in these mountains will probably be banned soon, as the human population grows and the need for clean water increases.

BIODIVERSITY THREAT:
Invasive Species

Alpine zones are usually remote and harsh environments, so it is difficult for invasive species to reach them and survive in them. However, small alpine zones are at much greater threat from invasive species.

Invasive Species of Australia's Alpine Zones

Australia's alpine zone is most threatened by invasive species. The total area of alpine zone is less than 31 square miles (80 sq km), and it is at a low altitude and has mild winters. These conditions allow introduced species to enter the zone easily and become invasive.

Hawkweeds are highly invasive and have become established in alpine zones in Australia and New Zealand, and in some regions of Canada and the United States.

Hawkweeds

Orange hawkweed and king devil hawkweed were brought to Australia from Europe as decorative garden plants. They have recently appeared in Australia's alpine zones, where they pose a serious threat to biodiversity. They grow densely between grass bunches, displacing many native alpine plants.

Hikers are helping government agencies to control hawkweed. When hikers come across hawkweed plants, they record the position and conservation officers move in later to spray the plants with herbicide or pull them out.

Fungus Threat

Invasive species can bring in diseases that affect native species. Chytrid fungus threatens the survival of the corroboree frog, an endangered species that lives in remote alpine bogs in southeastern Australia. Scientists believe this fungus may have been introduced by frogs brought from overseas as pets.

Brumbies

Brumbies are **feral** horses found in Australia. They are descended from packhorses that were released into the wild after European colonization. They trample alpine plants, which have little time to recover in the short growing season in summer. When brumbies drink from alpine bogs, they trample the sphagnum moss and they muddy the water. They also produce a lot of dung, which increases the nutrient content of the soil, allowing invasive weeds to grow. Brumbies can also carry weed seeds in their hoofs, manes, and tails.

Conservationists want to remove brumbies from alpine regions. Other people believe the brumbies are an important part of the history of these areas and just want their numbers to be controlled.

About two thousand brumbies live in the alpine zone of southeastern Australia. This invasive species tramples alpine plants.

BIODIVERSITY THREAT:
Climate Change

The world's average temperature is rising because levels of certain gases, such as carbon dioxide, are increasing in Earth's atmosphere. The increase in temperature is causing changes to Earth's climate. These changes threaten the biodiversity of alpine zones.

Climate Change In the Past

Climate change is a natural part of Earth's history. The habitats and biodiversity of different areas have changed as climates have changed, such as changing from rain forest habitat to dry woodland habitat.

Today, scientists believe the rate of climate change is faster than ever. Over the past one hundred years, many **glaciers** around the world have been shrinking at a rate of up to 30 feet (9 m) a year because of an increase in temperature. Scientists believe that many species will have too little time to adapt to climate change and they will become extinct. Other species will benefit from the changes.

As average temperatures rise, the snowy areas on some mountaintops are melting and alpine habitats are changing.

Mammoths of the Past

Rock carvings of woolly mammoths in the European Alps date back many thousands of years. This species died out in Europe between 10,000 and 12,000 years ago, because the climate became warmer and because of hunting by humans.

New Zealand's rock wren is threatened by climate change, because warmer temperatures may allow invasive species such as foxes and cats to enter alpine zones. The rock wren is a poor flier and would find it hard to escape these hunters.

Effects on Alpine Biodiversity

If average temperatures rise, alpine zones will shrink. The weather will be warmer at higher altitudes, so trees will be able to grow higher up on mountainsides. Forest will replace alpine vegetation and alpine animals will lose their habitats.

Many alpine areas are like islands of alpine habitat among a sea of other habitat types, such as forest habitats. As these "islands" become smaller, many alpine animals will have nowhere to go.

Invasive Species

Climate change may increase the number of invasive species in alpine zones. New Zealand's alpine zone has more than 600 plant species, and more than 550 of these are endemic species. Scientists used a computer model to predict what might happen in the extreme case of an average temperature rise of 5.4 degrees Fahrenheit (3 degrees Celsius). They found that half of the alpine plant species may disappear as invasive plant species, such as hawkweeds, grasses, and thistles, spread into New Zealand's alpine areas.

Alpine Conservation

Conservation is the protection, preservation, and wise use of resources. Conserving alpine biodiversity involves protecting it from threats caused by human activities. Research, education, laws, and breeding programs are very important to alpine conservation.

Research

Research studies are used to find out information, such as how alpine ecosystems work and how humans affect them. Research helps people work out ways of conserving alpine habitats. The people who carry out this research are usually scientists employed by governments, colleges, botanical gardens, zoos, or conservation organizations such as WWF (World Wildlife Fund).

Education

Educating people about alpine habitats is essential for alpine conservation. Information from scientists must be passed on to people, including students, farmers, and tourists. If people can see how alpine biodiversity is important, they are more likely to help conserve it.

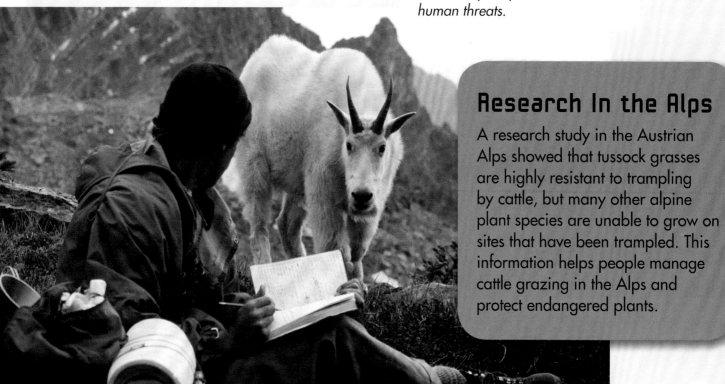

Research scientists study alpine animals, such as mountain goats, to find ways to protect them from human threats.

Research In the Alps

A research study in the Austrian Alps showed that tussock grasses are highly resistant to trampling by cattle, but many other alpine plant species are unable to grow on sites that have been trampled. This information helps people manage cattle grazing in the Alps and protect endangered plants.

Laws and Agreements

Laws and agreements are an important part of biodiversity conservation. One of the most important international agreements is CITES, the Convention on International Trade in Endangered Species of Wild Fauna and Flora. CITES aims to prevent endangered animal and plant species from being traded between one country and another.

The illegal trade in animals is worth about $160 billion a year, and huge amounts of money are paid for endangered species. Preventing the trade of alpine species is very difficult because these animals are poached in very remote places.

Breeding Programs

Some endangered alpine animals are bred in zoos. Zoos keep records of each animal to make sure that related animals do not breed. Some animals are moved between zoos so they can breed with unrelated animals. This helps keep the captive population of the species healthy.

Some snow leopards are bred in zoos, where they are safe from the threats that they face in the wild.

CASE STUDY: The Alps

The Alps are a large mountain system in western Europe. They are home to more than 20,000 animal species and about 14 million people.

Habitats of the Alps

The alpine zone of the Alps is between about 6,560 and 9,845 feet (2,000–3,000 m) above sea level. There are many types of habitats, such as bogs and alpine meadows, and different animal and plant species live in the different habitats.

Animal Species of the Alps

ANIMAL GROUP	APPROXIMATE NUMBER OF KNOWN SPECIES
Invertebrates	20,000
Birds	200
Fish	80
Mammals	80
Amphibians	21
Reptiles	15

Note: This table includes species from all habitats in the Alps, including habitats in the alpine zone.

Cliffs and Screes

Wallcreepers are birds that build their nests in rocky crevices in alpine cliffs and feed on insects and spiders. The alpine ibex climbs very steep cliffs in search of plants to eat. These goats move downhill to spend the winter in the shelter of the forests. Wildflowers such as saxifraga, valerian, and toadflax grow on screes, as well as in nooks in the cliffs.

Did You Know?

The word *alpine* comes from the name of the Alps mountain range in Europe. Some countries have mountain ranges named the Alps too, such as New Zealand's Southern Alps. Even a mountain range on the Moon, the Montes Alpes, was named after the European Alps.

The Alps cover about 74,130 square miles (192,000 sq km) across eight different countries in western Europe.

Alpine Meadows

Grasses and wildflowers, such as pansies and orchids, grow in alpine meadows. Flowers are plentiful in the springtime and attract many butterflies. Other insects appear too, and these insects attract birds. Alpine accentors, water pipits, and wheatears spend the winter in the valleys, but breed in the meadows. They build their nests among rocks or in alpine plants.

Grazing animals such as chamois visit the meadows in spring to graze on new growth, then spend winter in the shelter of the forests below. Marmots (similar to squirrels) live in alpine meadows all year round. These rodents graze over the spring and the summer, then hibernate in burrows over the winter.

Bogs

Plants such as sphagnum moss and sedges grow in places where water collects in sunken hollows, called bogs. These bogs are breeding places for mosquitoes and frogs.

Snow Line

At an altitude of about 9,850 feet (3,000 m), the ground is permanently covered in snow. This is called the snow line. Above the snow line, lichens grow on exposed rocks. There are few plants, because it is very cold and there is little if any soil for a plant's roots. The only animals seen here are golden eagles, which fly overhead in search of marmots and other prey.

Lanza's alpine salamander is an amphibian that is found in European alpine meadows at altitudes up to 8,530 feet (2,600 m).

CASE STUDY: The Alps

Threats to the Biodiversity of the Alps

There are about 14 million people living in the Alps region. Only 7 percent of these people live above 3,280 feet (1,000 m) altitude, but most of these people depend on the 100 million tourists who visit the Alps each year and who spend $60 billion annually. The ski industry greatly threatens alpine biodiversity, and climate change is another important threat.

The habitats of alpine flowers, such as edelweiss, are threatened by tourists and the ski industry.

Skiing

There are 300 ski areas in the Alps, covering about 1,313 square feet (3,400 sq km). The effects of skiing on thirty-three alpine bird species, including nine threatened species, were studied at seven sites in the Italian Alps. Scientists compared three habitat types: new ski runs, grasslands near ski runs, and grasslands well away from any ski runs. Not surprisingly, the study found that both bird species diversity and bird populations were highest in the grasslands that were well away from the ski runs, and lowest in the new ski runs.

Artificial snow is made at many ski resorts to make up for shortages of snow cover during the ski season. Chemicals are used to stabilize this snow, and the environmental effects of these chemicals are not yet known.

Climate Change

A shrinking alpine zone, caused by an increase in average temperatures, would affect many alpine plant species. Some alpine species will be able to survive by moving a little higher up the mountain, but species that live only in the upper alpine zone will have nowhere to go. Twenty-five alpine species have been identified as having medicinal uses. If these species were to become extinct, they could no longer be used to help people.

Biodiversity Conservation in the Alps

There are about 900 protected areas, including national parks, in the Alps. These protected areas are managed by different countries and cover about 25 percent of the total mountain area. Alpine biodiversity is linked to the biodiversity of other ecosystems, such as forests and wetlands. The Pan-European Ecological Network aims to link the different protected areas by conserving the areas of habitat between them.

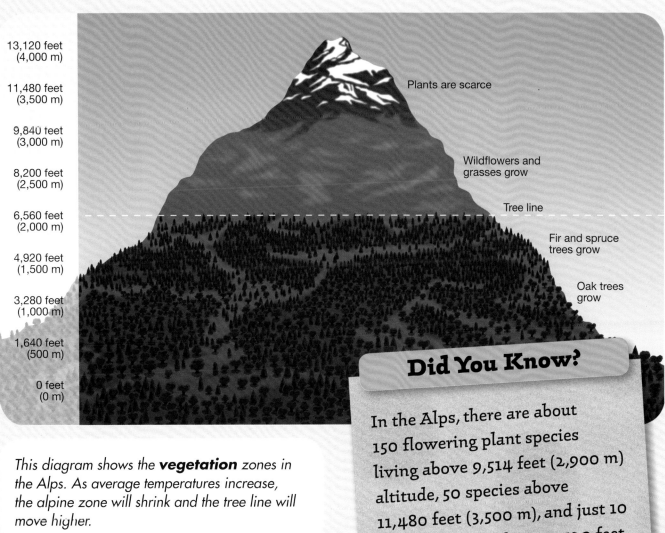

*This diagram shows the **vegetation** zones in the Alps. As average temperatures increase, the alpine zone will shrink and the tree line will move higher.*

Did You Know?

In the Alps, there are about 150 flowering plant species living above 9,514 feet (2,900 m) altitude, 50 species above 11,480 feet (3,500 m), and just 10 known species above 13,120 feet (4,000 m).

What Is the Future of Alpine Zones?

The biodiversity of alpine zones is under threat from human activities, such as tourism, farming and poaching. Climate change, caused by humans, is one of the greatest threats. Although the threats are large, concerned community groups and individuals can help.

What Can You Do for Alpine Regions?
You can help protect alpine zones in many ways.
- Find out about alpine zones. Why are they important and what threatens them?
- If you live in or near an alpine region, join a volunteer group and replant native plants.
- Become a responsible consumer, and do not litter.
- If you are concerned about alpine habitats in your area, write to or email your local newspaper, your state congressperson, or a local representative, and express your concerns. Know what you want to say, set out your argument, be sure of your facts, and ask for a reply.

Useful Websites

- **http://wwf.panda.org/what_we_do/where_we_work/alps/**
  This WWF website has information about the European Alps and conservation issues and projects.

- **www.biodiversityhotspots.org**
  This website has information about the richest and most threatened areas of biodiversity on Earth.

- **www.iucnredlist.org**
  The International Union for the Conservation of Nature (IUCN) Red List has information about threatened plant and animal species.

Glossary

adapt Change in order to survive.

altitude Height above sea level.

carbon dioxide A colorless and odorless gas produced by plants and animals.

climate The weather conditions in a certain region over a long period of time.

domestic Tame and kept or cultivated by humans.

ecosystem The living and nonliving things in a certain area and the interactions between them.

endemic species Species found only in a particular area.

erosion Wearing away of soil and rock by wind or water.

evolve Change over time.

extinct Having no living members.

feral Wild, especially domestic animals that have gone wild.

gene Segment of deoxyribonucleic acid (DNA) in the cells of a living thing, which determines its characteristics.

glacier A large piece of ice that moves slowly across the land.

habitat Place where animals, plants, or other living things live.

heritage Things we inherit and pass on to future generations.

hibernate Spend the winter in a dormant state (as if in a deep sleep) to conserve energy.

interaction Action that is taken together or actions that affect each other.

invasive species Nonnative species that negatively affect their new habitats.

migratory Moves from one place to another, especially seasonally.

nutrient Substances that are used by living things for growth.

organism Animal, plant, or other living thing.

pelt Animal skin with the fur or hair still on it.

poach Illegally hunt or capture wildlife.

predator Animal that kills and eats other animals.

range The area in which a species can be found.

scree Slope at the base of a cliff made up of the rubble that has fallen from the cliff.

species A group of animals, plants, or other living things that share the same characteristics and can breed with one another.

urbanize Change or develop into towns and cities.

vegetation Plants.

Index

A
Alps, 8, 17, 22, 24, 26–29, 30

B
biodiversity hotspots, 14, 15, 30
birds, 16, 23, 26, 27, 28
black-necked crane, 15
breeding, 6, 10, 15, 17, 24, 25, 27
brumbies, 21

C
carbon dioxide, 6, 22
climate change, 14, 15, 22–23, 28, 29, 30
CITES (Convention on International Trade in Endangered Species of Wild Fauna and Flora), 25
conservation, 17, 18, 20, 21, 24–25, 29, 30

E
ecosystem diversity, 4, 6–7, 12–13
ecosystems, 4, 6, 12–13, 24, 29
education, 24
endangered species, 7, 18, 19, 20, 24, 25
endemic species, 10, 11, 14, 15, 23
Ethiopian wolves, 18
extinct species, 5, 6, 22, 29

F
fish, 26
flowers, 4, 7, 10, 12, 13, 26, 27, 28, 29
food, 6, 12, 13, 15, 18
food chains, 12
food webs, 12

G
genetic diversity, 4, 6, 17
glaciers, 22
grazing, 9, 14, 15, 18–19, 24, 27

H
habitats, 4, 8, 9, 10, 11, 13, 15, 16, 17, 18, 22, 23, 24, 26, 28, 29, 30
hawkweeds, 20, 23
Himalayas, 6, 8, 10, 15, 18

I
inbreeding, 17
insects, 5, 13, 26, 27
invasive species, 14, 20–21, 23

L
laws, 24, 25

M
marmots, 12, 15, 27
medicines, 6, 15, 29
microhabitats, 4
mountain pygmy possum, 10

P
poaching, 14, 15, 18–19, 25, 30

R
research, 24
roads, 16, 17

S
skiing, 16, 28
snow leopards, 9, 15, 18, 19, 25
snow line, 8, 27
species diversity, 4, 5, 10, 11, 14, 15, 17, 23, 26, 27, 28, 29

T
threats to biodiversity, 5, 14–15, 16–17, 18–19, 20–21, 22–23, 24, 25, 28, 29, 30
Tibetan Plateau, 8, 9, 15
tourism, 14, 16–17, 24, 28, 30
tree line, 8, 9, 15, 29

W
websites, 30